D1203208

For McKinzie and Claudia.
—TS

Cover and internal design by Allison Sundstrom/Sourcebooks

The artwork was created using pen and ink, and colored in Photoshop.
Published by Sourcebooks eXplore, an imprint of Sourcebooks Kids
P.O. Box 4410, Naperville, Illinois 60567-4410
(630) 961-3900
sourcebookskids.com
Library of Congress Cataloging-in-Publication Data is on file with the publisher.
Source of Production: Leo Paper, Heshan City, Guangdong Province, China
Date of Production: March 2021
Run Number: 5021014
Printed and bound in China.
LEO 10 9 8 7 6 5 4 3 2 1

Turtles are found on every continent...

EXCEPT ANTARCTICA!

Todd Sturgell

This is a turtle.

Turtles are found on every continent except Antarctica.

They are cold-blooded and cannot survive in a cold, harsh...

Hey, where are you going?

You can't go to Antarctica.

Get back to your page, *this instant!*

Let's forget the turtle and move on.

Oh, look. Here's an owl.

Owls are found on every continent except Antarctica.

Hello there, owl.
This is a book of
animal facts.

Here's a fact:
Owls don't like to
be disturbed by
animal facts.

Owls rarely pay any attention to turtles!

Those two will never make it.

On to the next animal fact.

Dung beetles are found on every continent except Antarctica.

Hey, wait. Where's the dung beetle?

He's with us.

Why would a dung beetle travel to Antarctica
with a turtle and an owl?

This book is not going as planned.

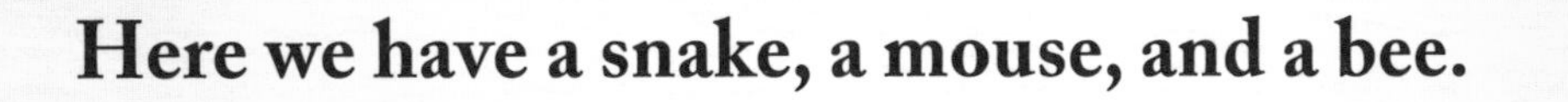

Here we have a snake, a mouse, and a bee.

Guess what? They're found on every continent except Antarctica—*and it will stay that way!*

Oh, you've got to be kidding me!

Frogs are found on every continent except Antarctica.

This one won't be distracted by a wandering group of rogue animals.

IGNORE THEM, FROG!

A turtle, owl, dung beetle, snake, mouse, bee, and frog would NEVER travel together on an expedition to the frozen continent. And they certainly couldn't cross an ocean to get there.
No boat. No Antarctica!
Now, can we please get back to normal?
I've got this.

This is ridiculous!

Pay no attention to these scoundrels! Turtles, owls, dung beetles, snakes, mice, bees, and frogs are still found on every continent EXCEPT ANTARCTICA.

Look at those strange penguins.

The coldest temperature ever recorded on Earth was in Antarctica.

(-128.6° Fahrenheit or -89.2° Celsius to be exact.)

Animals not accustomed to the howling wind and bitter cold air will find Antarctica unbearable.

Let's go home.
HA! Like I said, the Antarctic climate is too extreme for these animals.

Whew, I'm glad that's over.

As long as we're here,
let's talk about emperor penguins.

Emperor penguins are found
only in Antarctica.

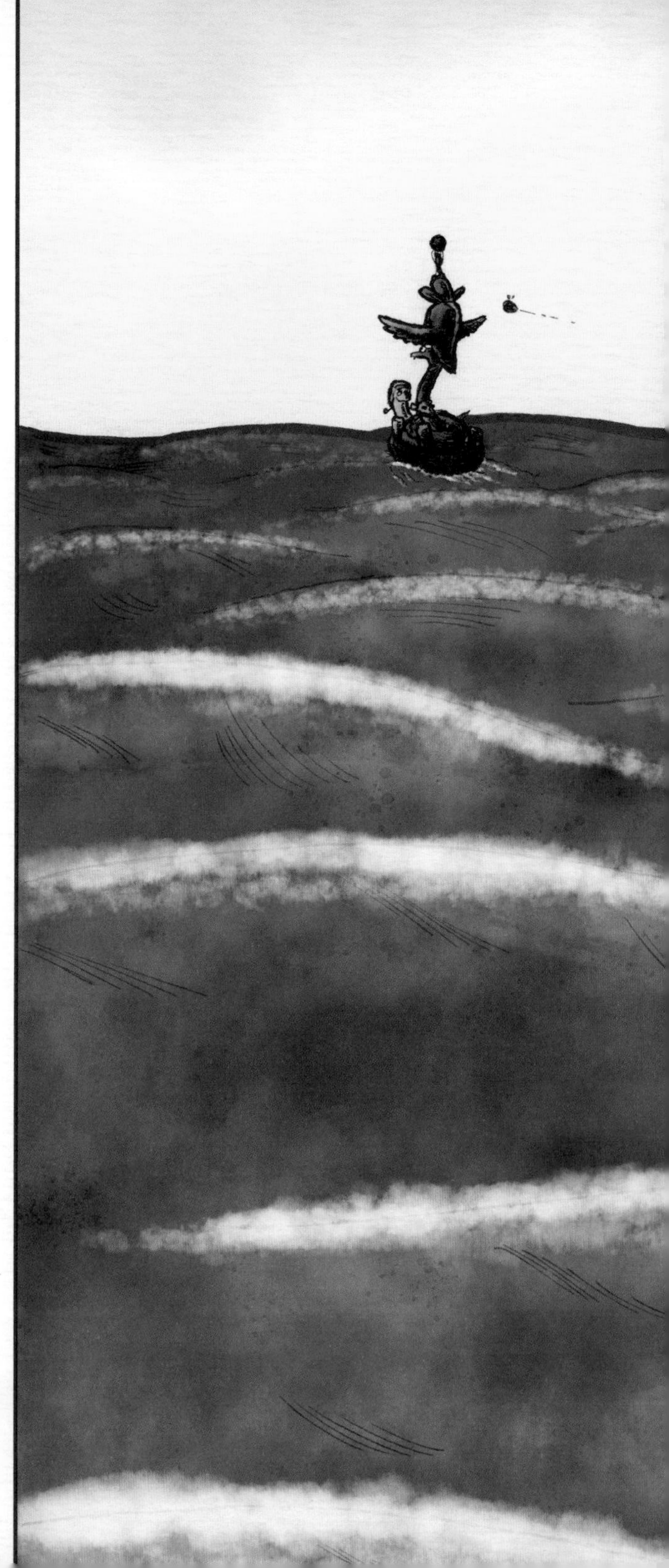

Oh no.

Not again.

TURTLE

- Turtles are identifiable by their distinctive shell.
- There are more than 356 turtle species!
- Turtles have been around since the time of the dinosaurs.
- The oldest turtle in the world was a 255-year-old tortoise named Adwaita.

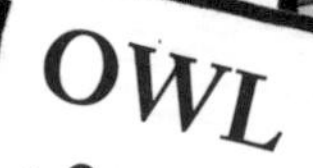

DUNG BEETLE

- Dung beetles navigate using the Milky Way!
- Some dung beetle species roll dung away, others tunnel underneath, and some just live inside the dung pile!
- Dung beetles fly, but not with their dung.
- Adult dung beetles don't eat dung, they just suck the moisture out of it. Mmmm, dung juice!

OWL

- Owls fly silently to sneak up on their prey.
- Owls don't poop! They eat their prey whole and barf up the undigested parts.
- Burrowing owls love to eat dung beetles. Uh-oh!
- Owls can turn their heads almost all the way around.

SNAKE

- Snakes can eat prey bigger than their heads.
- Snakes smell with their tongues.
- Snakes do not have external ears for hearing.
- Pythons are constrictor snakes with eighty teeth but no fangs.

MOUSE

- A mouse's tail can grow longer than its body.
- Mice eat fifteen to twenty times a day.
- Mice can squeeze through dime-size holes.
- Mice communicate through odor, body language, and sound.

BEE

- Bees harvest nectar from flowers to make honey.
- Bees dance to show the hive where to find food.
- Bee have a favorite color: blue.
- Bees create their own air conditioning by fanning the hive with their wings.

FROG

- Frogs have long back legs to help them leap.
- Male Darwin frogs swallow their tadpoles and later cough up little frogs.
- Frogs drink water through their skin.
- Frogs were the first land animals with vocal cords.

MORE ANIMAL FACTS

There are many kinds of animals that are found all over the world. Bats take to the night skies around the globe. You'll find hawks soaring through the skies above many countries. Spiders spin from north to south. Lizards crawl from east to west. And pigeons peck crumbs from Australia to Alaska.

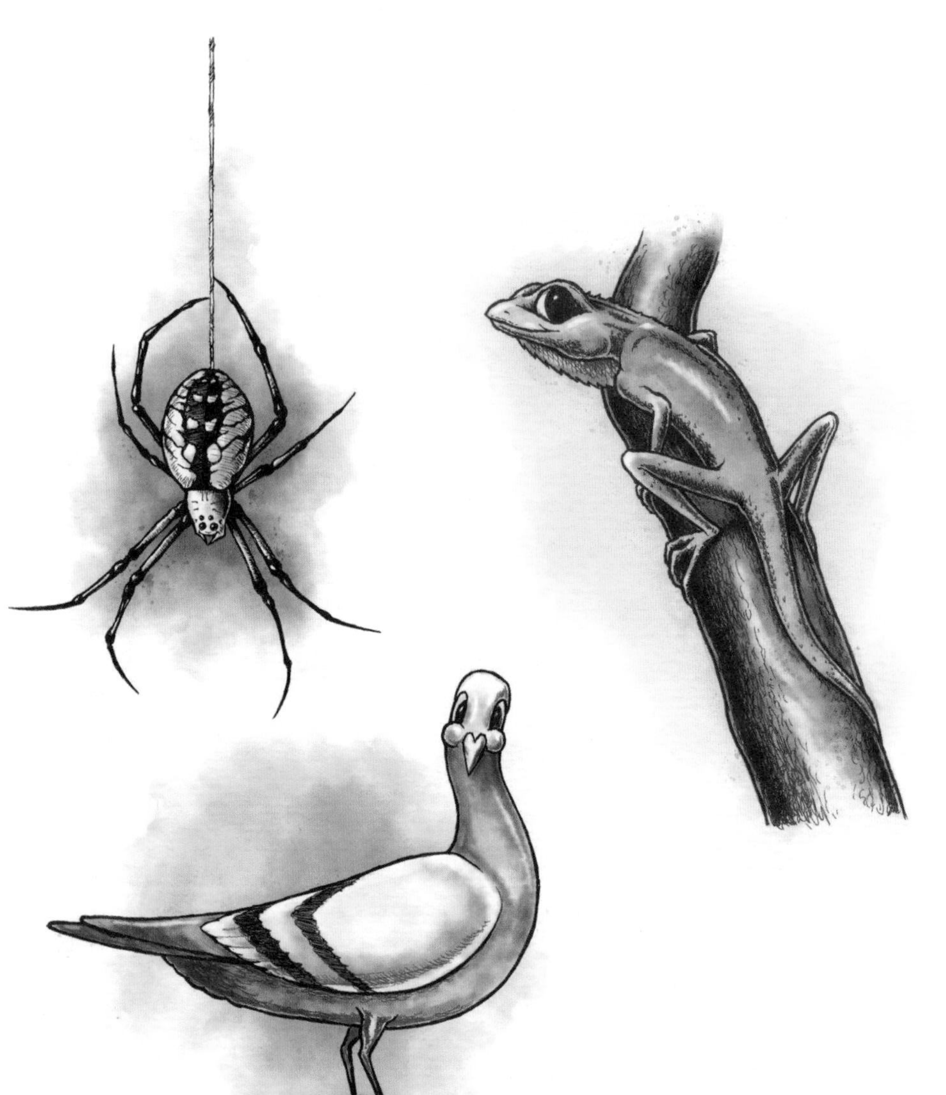

But there's one place where you won't find any of these animals. One place too cold for spinning webs and lurking lizards. That's right! You'll find bats, hawks, spiders, lizards, and pigeons on every continent...

EXCEPT ANTARCTICA!

ANIMALS OF ANTARCTICA

There are a surprising number of animal species that do call Antarctica (and the surrounding waters) home. The only native mammals you'll find on Antarctic shores are seals, but you'll find whales, dolphins, and porpoises swimming in the Southern Ocean. Of the eighteen species of penguin, only seven species live in Antarctica: Adélie, Chinstrap, Emperor, Gentoo, King, Macaroni, and Southern Rockhopper. Antarctica is also home to a wide range of seabirds, including albatrosses, petrels, gulls, and cormorants.

BRRR! THE FROZEN CONTINENT

Antarctica is very cold. In fact, it's freezing! That's why maps of Antarctica are mostly white—the land is almost completely covered in ice and snow. Antarctica is home to massive ice shelves, towering mountains, and a vast, flat plateau. It covers the south pole and is surrounded by stormy oceans. And even though it is extremely cold, Antarctica is also the world's largest desert because it gets the smallest amount of precipitation (rain or snow) every year!

A CONTINENT FOR SCIENCE

People have been visiting and studying Antarctica for many years, but no country can claim part of Antarctica as its own, though many countries have scientific bases there. The Antarctic Treaty Agreement set aside the entire continent as a scientific preserve. The treaty bans military activity and establishes freedom of scientific investigation. The treaty was signed in 1959 and entered into force on June 23, 1961.

KEY
Research Station
ANTARCTICA
France
Australia
United States
New Zealand
SOUTHERN OCEAN
INDIAN OCEAN
Russia
Ross Ice Shelf
Russia
South Pole
United States
Australia
Russia
China
Amery Ice Shelf
Australia
Ronne Ice Shelf
United Kingdom
Argentina
Larsen Ice Shelf
Argentina
United Kingdom
Ukraine
United States
Japan
SOUTHERN OCEAN
India
South Africa
Russia
Germany

CLIMATE CHANGE

When weather changes over a very long period of time, it's called climate change. Scientists study climate change to figure out what causes it, and whether or not we should do something about it.

Antarctica is still cold, but it's not as cold as it used to be. In fact, parts of Antarctica where a lot of the animals live is starting to melt a little bit at a time. Right now, scientists have observed that the world is getting warmer due to climate change and have discovered that some of the ways people create energy has caused the problem. This is called global warming.

WHO TURNED UP THE HEAT?

When the world gets warmer, animals that are used to living in the cold have to work harder to find food and raise their young. When the ice melts, it goes into the oceans and the oceans can rise. People who live near the water don't want the water to rise up to their homes.

This may seem scary, but there's good news!

Scientists and leaders all over the world are working to fix the problem and they've come up with solutions that will help. They are creating energy that doesn't make the world hotter. We call this clean energy.

YOU CAN HELP TOO!

People just like you can help too by looking for ways to make the world a better place for the people, animals, and plants that live and grow where you live. Planting trees or starting a vegetable garden can help clean the air.

Recycling reduces waste. Turning off the lights when you leave a room saves energy. Parents and teachers can help you find even more ideas so ask them to help you find ways to reduce, reuse, and recycle.

There are so many people working to fix the effects of climate change that I'm sure when you decide to visit Antarctica like Turtle did, you'll find it's still a perfect place for penguins.

GLOSSARY

Antarctica — Earth's southernmost continent. It is a cold, ice-covered land mass and includes the south pole.

Clean energy — Energy produced without making pollutants and gasses that can contribute to climate change.

Climate — Average weather conditions for a particular area or time.

Climate change — A change in weather conditions over a long period of time.

Cold-blooded animals — Animals with body temperatures that change depending on the environment. Also called poikilotherms. (That's fun to say.)

Continent — Any of the world's large divisions of land.

Global warming — Climate change that is caused by people and that makes the world warmer.

Ice shelf — A floating piece of ice that is attached to the land.

Plateau — An area of level, high ground.

South pole — The southernmost spot on the Earth.

Temperature — How hot or cold something is (like the air outside) as measured on a scale.

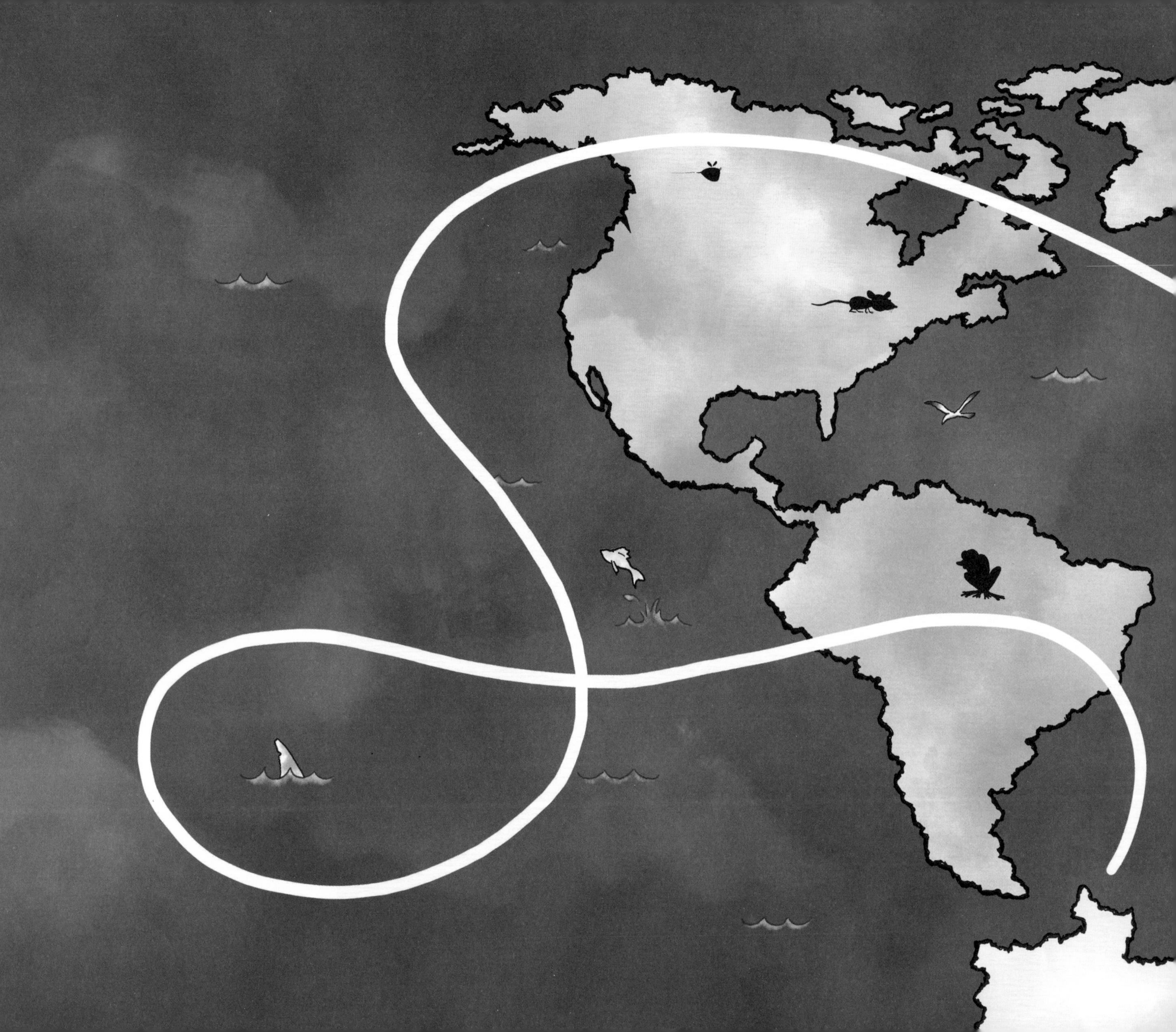